Nothing is Impossible
M. Light Shields

To Gale, Kim, Mary, and Sherry

Contents

Introduction

Since the dawn of time, human beings have sought to comprehend and document their beginnings, and quantify their purpose In the Universe. The earliest cultures all developed rich archetypal mythologies that told, not only the stories of their beginnings, but quantified most of the energies and environmental phenomena, personalizing the experience into beings. To early humans the world was alive. Their God's breathed life into their world and gave them a sense of purpose in the universe.

As time passed, many of these metaphors were replaced with discoveries, and as years of discoveries added up, it would dawn new conclusions, which would lead to systems and theories, and laws. Some of these concepts have acquired certain features, that were never really part of the original data, but existed beneath the surface as culture. Similar to the way that archetypal humans filled their world with living things, these beliefs around certain prescribed perspectives set the stage for various degrees of possible miscalculations in the way that scientists observed each new discovery. Ironically, history has also taught us that perception, coupled with our relationship to information, can and will change over time... Again and again.

If we go back to the first atomic model, we can see exactly how the physical structure of the universe was applied to it.[1] Down the road, most of the science community, seemingly in agreement with this unspoken rule, held true to the culture of form. Even the naming schema for atoms gives us particles. These were originally thought to be the proton, electron, and the neutron. And, it is still believed today that, the electrons are scattered about the nucleus in a "sort of cloud" (science encyclopedia). Ironically, no one seems to think this a little unscientific, but rather see it as an unspoken agreement that, there are things we don't understand well enough to explain precisely. And it seems to be a solid feature of the cultural fabric of modern science.

The goal of this book is to break down the barriers that culture has placed upon the current view of the universe and replace it with a totally new perspective. One with a learning state and a dynamic information stream. As we discuss these ideas, it may be necessary at times to literally deconstruct the old concepts, in order to reexamine them. This can challenge the very nature, of the fabric, we understand to be reality, not only of our self, but as well as the completeness and depth of the known universe. This perspective, we hold, must be affected directly, in order to be shifted, into a new perspective. It is the nature of information. It has been the same since it first was, that our perception, over time, has always been the medium of change. And, perception/time = state of total mass.

Part 1

Deconstructing the Myth

Chapter One

The Big Bang

In the early 20th century, a couple of guys used a radio receiver to measure the cosmic background radiation in space. One of the things they discovered was a field of gamma radiation that appeared to be equal in density no matter which direction they pointed the receiver. From this data, they came to the conclusion that this field of radiation must be a remnant of a huge explosion, which gave birth to the known universe. And, to me this seems like a guess. [1][2]

The theory is this, that in the beginning there was a singularity, meaning all time and space, which conveniently already existed, were packed into a "non-spacial" state and released into the void simultaneously, with a big bang. The theory, now expanded by others, continues by stating that when entropy has run its full course, and when the momentum of expansion could no longer fight the resistance of gravity, that the whole universe full of matter would be pulled together into another singularity, where the whole process would start all over from the beginning.

In no time at all, it would crash into a few natural obstacles. These, minor flaws, would of course be solved with newer super "unprovable" data. The first problem they found was that the volume of mass was actually picking up speed and would therefore be beyond the power of the existing state of total gravity to ever stop its expansion. This problem was solved with even more (theoretical) mass. It has to be there if the big bang is true, right?! So, scientists came up with a solution, and announced it as something... Invisible. Something they call "dark matter", and they have spent this whole time, trying to find it. I guess it's difficult to find due to the fact that its dark.[4][5]

The next problem they found, at least important enough to be considered an issue worth solving, was gravity, which seems to break down at the quantum level. And frankly if there was enough gravity to pull the entire mass of the universe into a singularity it would constitute an important factor IF...! The theory was accurate. Otherwise it would be natural that the information and the system don't align.[6]

The problem is that it may not be accurate at all. In fact, it is likely that we have overlooked a great many details, in order to protect this historic relic of scientific culture. It is also natural for most people to be influenced into believing a theory purely as a cultural habit. Especially one so steeped in our daily life.

Lastly, the theory itself seems like an odd guess at explaining the phenomena correctly. If there was an explosion big enough to thrust out an entire universe, it would have a huge collective hot spot in the place where it occurred. At that location (directionally from Earth) there should be a massive density of gamma radiation, while the opposite direction (away from the center) there would be almost none, and the perpendicular angles would be equal to each other.

There are recent discoveries, as well, that don't fit this model correctly. Scientists in Hawaii [7], using infra-red to measure the distance of galaxies, discovered they could measure them in deep-deep space. The furthest so far being 10-12 billion light years away and it's old. This gives a substantial fracture to the current model because it isn't just saying that the galaxy they found, at that range, is 10-12 billion years old. It is actually saying that, an aging galaxy existed at this bearing and depth approximately 10 billion years ago, placing the Universe well past the 13-15 billion year range, the Big Bang is based on. The fact is that if the Big Bang theory is missing all these specific details, that would seem to validate it's truth, what really did happen to give birth to such an expansive system. It might be necessary to build an entirely new model, in order to accurately resolve this snap shot. One without gravity, but...

Our collective reasoning, when it comes to systems is usually pretty cut and dry for most of us. If it ain't broke don't fix it. But, what if it is..?

Chapter Two

Heisenberg Uncertainty Principle

One of the founders of Quantum Mechanics, Werner Heisenberg, observing the movement of electrons, discovered that the electron didn't move the way a form was expected to behave. Sometimes it could be observed in a specific location, but whenever this would occur, the vector of the electrons path could not be determined. Other times the case would be reversed. In other cases, what was known to be a single electron, could suddenly appear on two separate shells, and then return to one.[8]

Heisenberg decided that there was no way observe the electrons position and vector at the same time and referred to this as the Uncertainty Principle. While it may very well be the case that these details are persistent agreements that phenomena may yield, my question is; is it wise to frame the perspective as a "principle", when it may be the model that is unable to explain the phenomena, and therefore the model itself that needs rethinking.

If we look at the argument itself, we can see that it is framed in the *language* of form. It seems to assume that if we look closely enough, the atom's electrons will appear in a location and maintain a direction. This entire model is based on objects occupying a field of space, that volume of space filled by an object circling another object, and its wrong. This point of view was catalyzed by the current cultural theme, which is again based on form.

The electron is not actually a form at all. In fact when we "view" it in a spacial relationship as mass, we are actually misinterpreting the information. While I agree we are seeing the appearance of form, it merely represents it to the observer, but let's try something different. Let's try looking at the same information, but this time with new angles, to see if we can do a better jobs of understanding the phenomena.

The Double Slit Test

Ironically, in 1803, Thomas Young, an English mathematician, performed the famous double-slit experiment that he later described in a paper entitled, "On the nature of light and colours". His accomplishment helped influence the light wave theory. He performed the experiment to observe the flight path of an electron fired from an electron gun. The electron fired from the gun was aimed at a barrier with two slits, with a recording plate behind it, that would record the event, (where the electron hit it).[9]

The first thing that was observed was that there seemed to be no way to predict where the electron would contact the recording plate, nor which slit it would pass through. But, the really significant discovery would actually happen when no one was watching the event. It seemed that when the observer left the experiment to run, the plate recorded something unexpected. Instead of the dots, left by the impacting electrons, they came back to find what appeared to be the impression of waves. The markings also had the inference wave of the harmonics between the two streams (from the two slits), colliding into each other on route.

This is really saying something quite significant about these so called "particles"! What it seems to be saying is that the electrons may only appear to be forms, when in truth they are not actually form at all. So, if the electron is a wave, that only appears as a form, then what about he atom?

The Boze-Einstein Condensate

In 1999, at Los Alamos Labs in New Mexico, another group of scientists set out to confirm the Bose-Einstein Condensate[10]. This theory, originally developed by Bose's observation of the light particle/wave, was extended by Einstein. What he found surprised him so much so that he did the math over to double check the outcome. It appeared based on the math that the atom itself could also be a wave.

The experiment used an advanced cooling chamber to reduce the temperature of a cloud of atoms to close to absolute zero. As the temperature cooled, the entire cloud was reduced to a single atomic state. In other words there appeared to be only one single "atom". I say atom, but how can millions of atoms appear as though there is only one. It would appear that they were capable of occupying the same space. Quirky yes, but not beyond explanation. There is one more thing to consider here, and that is, that the experiment would only work if there was no visible light, and no observer watching. Sound familiar.

In the slit test, without the observer we had, not a form, but a wave. Similarly, the Bose-Einstein Condensate gave us a cloud of mass, not compressed into a singularity, but rather by removing the observer and visible light allowed the truth to be seen more clearly; that the forms of these particles are more likely images that are animated into 3-dimensional objects when perception collides with the information packed in the wave.

There's something else the Heisenberg Principle tells us as well. If we locate the position of the electron, we lose the value of the vector, just like looking at a single frame of a movie, of a ball thrown into space. We only see the position of the ball within each frame, but the vector is lost until we add a second frame and form a relationship to the position of the ball in the first frame. This is exactly what Heisenberg has given us, the frames, but these frames are passing at 10E43/sec (that's a 10 with 43 zeros per second). It also states that the medium, just like the film, has nothing to do with the events it supports.

Chapter Three

Perception

In the early days of human beings, the world was alive. It was filled with living gods that represented the wonders of nature, like the Sun and the Rain. In some cultures it was believed that good tidings would fall upon those who did not upset the gods, otherwise your entire clan could be wiped out by unfavorable conditions. This same basic nature of thought has always underlined the surface of our explorations into the unknown. It is a core survival mythology that is based on real loss and suffering, whenever conditions would turn poor. Groups could have to migrate huge distances, possibly through rival territories, risking what little they may have had left. Don't upset the gods!

Later on, once human's managed to harness fire on the other hand, it unleashed a major shift in perception. Once we believed that these powers were limited to the gods alone, but now we possessed the same power as the gods. Looking back we can see that it was not only a shift in consciousness, but also a major shift in understanding the nature and technology of fire. Fire eventually leads to all the rest of the technology that follows. The message it holds is that someday we may possess all of the same powers, once limited to the gods. The first stick burning in their hands must have felt priceless. Like a really, really, rare jewel. This was "worth" protecting.

Throughout the history to follow, human beings would be faced with a shift in perception occurring again and again. Each new discovery adjusting the template of the world and sharpening the values into better new models. New models that can be applied more efficiently to existing problems than older models once could will eventually replace the other over time. It is the nature of perception to organize itself into higher states. We see it, not just in humans, but in everything that lives. It would be safe to conclude that one of the very natures of the universe is to "organize matter into higher states of perception", that this is not a "rule", but a rhythm.

We've talked a bit about perception, but what are we really talking about? We know that an observer perceiving events is using perception, and we know that everything living has some form of perception, but how do we define it? What is it, exactly, that makes it different, from say, a computer or a rock? One major difference is that a computer needs instructions before it can manipulate data. Where as it may have the capacity to reason, it must receive the initial directive in order to take action. And, its insight is limited to its process. In other words, it has no initiative. No impulse for survival whatsoever.

Perception, on the other hand, can reason. And not only can it reason, but it can deduce the truth. This statement is likely to turn some heads toward a moral standard of sorts, before really swallowing this pill, but think about it. Things like, $0 + 1 = 1$, and statements like "infinity can not be measured" resonate a true value. The symbols, of course, carry little meaning as (1) in the first axiom could be any symbol that represents a single instance. The point is to simply show that there is some elemental basis for logic that perception expresses as a constant. It is the thing that makes it true. To recognize when it is in alignment with the contextual value. In other words "true".

The Culture of Form

Around 100 bc, poets and philosophers had the idea of clusters of atoms, in reference to the early notion of chemical bonds. Chemical bonds are the strong and weak forces that bind atoms into molecules and pair elements. It would seem quite natural for people to consider their own nature and state when applying constructs to the new things they are discovering. And on this predication, it would also seem natural for scientists to assume that everything was made of "something".[11]

This is the template preception entered into the atomic age with. As scientist discussed the wobbly world of quantum mechanics, they could never really let go of the culture of form, that had been so cleverly injected into the perspective of everything we look at. As human beings, we have a tendency to follow the groove, and whenever we approach a new perspective and we usually observe it through the template we're familiar with. "Horse-power" for instance is a measure of an internal combustion engine's torque, while "thrust" is a jet engines force. Thrust once meant, the act of throwing a javelin. It's a throwback for sure.

This is exactly how we began to view the atom, in the beginning when Amedeo Avogadro gave us his molecular model. Later, we were so steeped in form that when we finally did reach the stage of rich quantum exploration we still really didn't understand the slit experiment at all or what it was saying. We decided that these "mysteries" were things that just couldn't be explained and trained ourselves to look past them.[12]

This same culture of form gave us the look into our own beginnings with the Big Bang Theory. The notion that everything can be stuffed into a singularity, or that the mass that occupies the Universe, has always existed are a bit unpalatable for someone who dines on truth. Call it a bad after taste, but whatever it is they keep using to correct all the problems it has, it still always seems to be missing something.

The Impossible Dream

So let's just look at the very idea of a singularity and ask some basic questions about it. The first problem I see, in light of these last few thoughts, is the lack of any kind of observer. If there was nothing left but a single collapsing black hole of mass, it could **not** be mass, it would be waves of information, as we saw in the Slit Experiment. We also have a heat problem that can not be solved; As we saw in the Bose-Einstein Condensate, the temperature was cold, allowing waves to stack, or more likely just hold the information in a neutral state.

But, just for fun. Let's try removing the unsavory conditions; the entire universe of mass, the missing needed gravity to cycle the event, the heat problem, and the failure to define the beginning at all. Oh, and the singularity, which leaves us with "Nothing".

This I agree with, that Nothing would have to be the state of a universe at some point before it could ever be anything. This I know has to be true. And, this becomes a context issue. What would nothing be and where could this possibly occur, independently and without an observer, which gives us a logical problem, that Nothing, without an observer is impossible. And, what is the motivation for a beginning if the entire thing has to dissipate all it's energy in order to become a singularity. What's the force behind the Bang.

It's going to feel quite natural to dispute these points. And yet the arguments won't solve any of the conditions we had to throw out just to get to the outcome, which itself is impossible as well. What is more true, that we check our Facebook on average up to five times a day or that all the mass in the universe can be stuffed into nowhere and recycled at will. And, based on pure logic alone, even Nothing is impossible, without an observer.

Ironically, all of these problems can all be addressed with a different cosmology. One that is based on perception, rather than form as the medium. One that is not based on a concentric models of forms, but based on logical processes, that follow an infinitely expanding theme... To organize matter into higher states of perception.

Part Two

Alice in Wonderland

Chapter Four

In the Beginning

Now that we've broken down the old existing model we can begin to build a more efficient model that not only fits with the existing science, but is more efficient at explaining the phenomena of the universe. The model we will be constructing I call the Single Wave Theory. The model we will be using will be broken down into components so we can discuss the finer points of each as I go over them. The first most important part, at least for us, is the plasmic field. This is the relative field of perception and the foundational concepts and boundries are defined by Einstein's Relativity. His math describes a sort of wrapper to our perception of the universe and the relative value (relationship) of energy, mass, and the speed of light.

The second part is how the actual frames of perception are forged; a combination of states that effect the plasma and the existing information about the known universe (to itself). In other words, Einstein's General Relativity attributes the glue of the universe to be gravity; that this holds the pieces together, but this is still a form based notion. The fact is that the information "about" the universe, which occupies no space, is all that is needed to prescribe the environment to perception, in order for experience to occur. Emmanual Kant describes this in his Critique of Pure Reason.

The last part would be the record of information "about" the universe; how we create it and experience it, as well as how we collect and store it. This is, as well, a notion of Kant's called the A-Priori. If Einstein is correct (and I believe he is), the universe is "in and of" itself, and therefor of it's own volition, which means that the information about it is retained by the whole collectively. This information is also not limited to time/space when it comes to componants sharing information about the global state. In other words, every state of the universe is passed to the next, along with a progressive accumulation of its momentum, like throwing a base ball.

So, in the beginning, there must have been a state of perception, which is required in order to have an event (according to our model), but what kind of perception would we be talking about. It would have to be somehow visual, if we are to consider talking about the physics, while it is likely without eyes. And, if we are really talking about a "true" first perceptual state, we must consider the value of how it might actually be perceiving whatever event or environment it is experiencing and observing, also how it might relate to itself and that environment. If we could even call it that.

If it is perception, and qualified to be considered perception, it is likely going to start at the most, absolute, null state. In other words the context is a huge factor in its cognizance; having no frame of reference to draw from, it would only have the existence of nothing to reflect on, the first logical problem to face.

Another thing we must consider is that this state of perception must have logically been in this state indefinitely before any change of state could occur. In other words, perception must have been infinite, before any following change of state. And this is due entirely to the context. It could not have been placed there, as Einstein realized. The universe must have come from itself. It would have corrupted any perceptual knowledge of a self and nothing, to know that it was being put into nothing; harmonically. It simply must have always been.

This point makes it very clear that the "state" of perception we are discussing is not an aware state, but rather an unconscious state. The difference being that a conscious state would be one in which it was aware of itself and the context being observed and here we have no indication there could be any of this in this first of states.

Chapter Five

Plank's Moment

At some "point", resolving the context to be null or non existant, ie., "nothing", becomes the first data to discover. Perception, having concluded that "this is nothing", ended it. Now it knows. In other words, perception aligning with the context would be like a pair of tuning forks, and the recognition of the state would respond directly to the energy of the perception aligned with it, thus causing a harmonic resonant response to the pairing. It would never be able to return to its pre-stated condition, ever again. It could never... un-know.

This harmonic, I believe, would begin to resonate a single wave. This wave holds a snapshot of the state in information, which must measure Plank's Constant, if we are talking about *this* universe, and would therefor represent a single quantitative state, or a quanta. I also believe that it returns a value, relative to the perception (this one the universe) observing the state. In this case, "true". And I believe that the entire process takes 1/10e43 of a second. I also believe that if there was a human observer (in space), with the ability to look at this state, it would see a single hydrogen atom. The reason I feel this way is that it is likely that the information equals the state based on the context of the perception observing it, and therefore holds all the information about the moment and its meaning in a metaphorical reference. In this case, an environment with a single element, which in this universe would be the simplest form... an atom[14]

The Tao of True

When perception, finally, aligns with the context, as in the cognizance of "nothing", it resonates a harmonic, being "true". Again, not in a morally implicative way, but rather that of the harmonic resonant value, like a perfectly tuned wheel is true. It would almost be like a reward. In fact, for a perception that has been engaged in nothing for eternity, this would be like crack. It would likely repeat this indefinitely, if the reward continued to respond to the realization.

This is a really interesting point when you think about it. It's the basis for all logic. The ability to recognize when a value is true, based on reduction. Perception evaluated nothing long enough to either accept, or learn to conceive of it as the value of the context, and that it was the only value to discover in this relative state. And once done, could never be undone. This moment has the capacity to become an entire singular node of science. Within it are the seed catalyst for all other impressions of perception, and this tells us things well beyond the basic logic.

Chapter Six

The Single Wave

For a moment, while Alice is tumbling down the rabbit's hole, she reflects on the lack of context, in which to form a relative position in space. She is unable to gauge her speed of descent nor her duration of falling. For Alice this problem is a simple lack of contrast. Without being able to see any kind of reference, to her momentum, she is unable to judge any changes in her own state, but she recognizes the issue as contextual.

With perception in our model this problem would be compounded by not only the missing context, but the lack of faculties in which to measure any by, if they did exist. In other words it would have no eyes to see what was missing. It would also have no context to frame the references needed to judge distance or rate of change. And lastly, it would have no clue that it was even in this predicament until it could cognize that it was evaluating anything at all. Now let's jump forward (an eternity) to the initial "true state" of the single wave and step backwards to see if we can unlock the possible steps that might be logical; that combine to equate to the sum of "nothing".

The first thing required for this logical statement to be complete would be the actual context of "nothing".

Ironically this doesn't need to actually exist. In fact if it's nothing, it really doesn't. Perception is merely aligned with the fact that it is observing nothing, which sparks the harmonic. This would appear to be a solution for a long term condition. It's likely that perception observed this fact infinitely before resolving that the context was missing. It also points out that there was a build up to this cognition, and that this process of searching ended when the solution presented itself.

The reason I say this must have taken a long "time", is to suggest a time-line already exists. In fact if we look at the moment just after the single wave it gives us a glimpse of that value. The value being the duration of time it takes to complete a full state of quanta, called Planck's Moment. And at some point logic would have had to make the jump from "looking for", to "realizing", nothing. Otherwise it would never have been able to escape the infinite loop of never finding any sort of answer.

The clock speed, for this process, could have been determined by the process speed of each query, as the duration through states. It could have been the same speed from the first, or it may have changed to the quantum rate once the single wave appeared. For now this is uncertain, but fortunately, the information still exists within the initial single wave itself. The **single wave** represents everything about what proceeded it, as it is the result of the pre-initial state. It is also likely to be the same now, as it would have been when it originally occurred, like a recording. And, it should be quite easy to find, as it is within every state to follow.

The Locigal Infinite

And again, logic must make another jump. This jump will cause what I believe to be the Big Wave. It may have been the case that the value of the initial single wave was being recreated over and over at Planck Speed, or there was a shift in logic, it uncertain, but in either case, the result would be an exponential explosion of Planck's Moments, each resulting in a single compound multiple; like 2+2=4, 4+4=8, etc... and if this were hydrogen suddenly coming into being at an exponantial rate. The gravity, and insuing heat and fluidity would only need a tiny spark to ignite. In one second, the volume of hydrogen would grow to 2^10E43, in one minute (2^10E43*60), in one hour (2^10E43*60*60), etc... It would fill the void with infinite possibillities.

And if nothing is impossible, anything is...

Part Three

Through the Looking Glass

Chapter Seven

Relativity and Perception

We were born into a world of objects, colors, and sounds, a world of change and dimension. We are comfortable with energy, relating to it as a component of our environment. We have mastered internal combustion, microwave technology, and found the Higg's Boson. Through technological advances, we have come to understand a great many things about our environment, and our culture. To most of us, we live in a world made of small building blocks that come together to create larger and more efficient objects.[15]

In the early days of science, everything was made of some kind of form, but new things came along that really began to change the way we viewed and operated within our environment. Consider the technological advances that radio, or cablegrams, gave us. We could now compress time as a service. Before this, the fastest way across the country was the pony express. Soon, with the invention of the telephone, it would place people inches from each other. All of which is happening in the world of quantum mechanics. Electromagnetic waves, piercing space to deliver information etched into the signal.

But let's back up and ask ourselves a question. Why didn't we put all this together when we had Einstein's brain on the block. Well, because Einstein and others, I believe, felt that quantum mechanics was a threat to physics, he considered incomplete. I know it was certainly missing to many important factors at the time to be taken seriously, but if he could have been convinced by the right tide, his input and influence would have propelled it into much richer depths, much earlier than it is.[18][19]

In reality, quantum mechanics was never a threat to physics. It was an entirely new layer, with an entirely unique precipice. It showed us that there are two sides to the universe, that both operate on entirely different rules. These two worlds, so to speak, must interact through some type of interface, and form a strong relationship to the other, in some way, shape or form.

To most of us, we expect that the forms we are familiar with will remain that way, which is likely to bring up the naysayers in almost anyone. Who really wants to be told that the world is a holo-deck, on a science fiction television show. It sort of robs the mind of the sense of purpose we create living in it. Makes it strange to think of it as fabric, much less a two-dimensional fabric.

Relativity

In the early part of the 20th Century, Albert Einstein gave us a formula for the events we perceive in the world around us, called, The Theory of Relativity.[17] His Theory was based on two premises; the speed of light and the conversion from energy to mass or $E = mc^2$.

In my opinion Einstein's Relativity is the "scope" for perception and gives us specific meters in which to measure it by. In a way He shows us the wrapper or the envelope that reveals the upper limits of perception; that being the speed of light.

One of the thought arguments he give us is the case of the car moving at the speed of light. He places two unique observers for the event in two unique locations. The first is "in" the car while the "other" is stationary, observing the car passing by, and what they are observing is the moment the headlights come on. Ironically, both observers see something entirely unique to their perception of the event.

The first (in the car) sees it as anyone would, and the wash floods the view ahead. In other words the lights come on like normal. The other (stationary) sees no sign of any headlight wash, as the car has already reached the upper limit of perception. And if this is true, it is describing something else, that the information about the universe is relative to the perception observing it, which means that perception is observing a unique universe.

It is rendered in the fact that there would not be a beam of light to see from a stationary position unless it could travel faster than the frames, while in the car the observer sees as if they aren't moving. Otherwise the light would not appear for them either, again due to speed of travel plus(+) the speed of light, but the math says otherwise.

Chapter Eight

The Universal Map

In the peak of the physics heyday, a small band of renegade scientists took off in the wake of this quirky new science called Quantum Mechanics. One of many scientists, who worked on collective pieces of the puzzle, came to see a need for a better model for physics. He and another scientist, this one a neurophysiologist, both separately developed different models based on the same basic concept; The holograph. I believe that this is how the information embedded in the wave is translated into dimensional reality.[19]

Let's go back for a second and consider our original single wave. The wave, itself, represents a single state in an infinite field. Now, we need to look at the wave, uninterrupted by the observer. What we should see, based on our query and interest, is the state of a single atom of hydrogen. It would appear, as form, only when being observed, so we must remove the observer, in order to see the wave itself. The information embedded within the wave, must have the information about the entire state of mass and all previous states. In other words, **the state is the map**. Within it are the spacial relationships to the whole; all adjacent states and the global states. It would also be subject to any change in state that follows, therefore, it is likely that the state of information is not subject to the speed of light. It is also likely that the state is not a secret to perception alone.

A. Priori

A man, by the name of Emmanuel Kant, came to the conclusion that the perception of each individual person, must have a unique access to a global map prior to experiencing it. To him this solved a problem within the mind, and how it was able to recognize the environment, after an event like blinking, or sneezing. How was the mind able to hold the information, between states. His conclusion was that, we somehow store a local instance of the current state of the environment in cache, so to speak, which can be called to refresh the state of the environment any time it's needed.[21] This way it wouldn't be like waking up in a strange house every time we sneezed.

Another man, named Dr. Pilbram was attempting to solve a similar issue, with the way the brain gathered, stored, and distributed information within the brain. His solution was a holographic template, where information could be stored and accessed throughout the brain, in wave patterns. An old friend, who painted Corvettes, helped me to see this a little clearer. I'm sure he wasn't aware of it, but he gave me a look at how the mind collected and stored information within the body.

His name was Jim, and he was replacing a single fender on a beautiful, 1974 metallic root beer Corvette, that was otherwise in mint condition. He was painting the whole car and not just the fender. So, I asked him why, and the answer really surprised me. He said it would be impossible to recreate the same atmospheric conditions, as when it was originally painted. This went way beyond the car, when he said it. I saw how the location, altitude, and magnetic fields would effect the metal flake, in the paint, once it hit the surface. The plates might not lay the same way, in the medium, and the fender wouldn't match the rest of the car.

I saw how this same process was occurring in the cells of the body. Our physiology incorporates metals into our cells, for different purposes; zinc, magnesium, and iron, just to name a few. I realized that we probably store memories in complete states, and not in specific areas of the brain, but throughout the entire body. This is, in essence, how we acquire the map, and when we reflect on it, we are literally embodying the entire state, which was written into the cells at the moment they were created, like a file packed with data about the event. Or, a root beer brown Corvette you'll never forget.

Mirror Neurons

I recently saw a video on YouTube done by Jeremy Rifkin, called Empathic Civilization. In this video he talks about an experiment done in Parma, Italy, studying the brain of a monkey to see which parts of the brain fired when he tried to open a nut. He explains that, during the experiment, a scientist, walked into the lab, and snacked on one of the nuts. The monkey, still wired for progress, was showing the same areas of the brain firing, as if he was himself opening the nut.[22]

The idea they concluded was that, the monkey was being inspired, by what they suggest to be something called mirror neurons, but I would beg to differ. I believe that the monkey is embodying the state of its own experience, that was written into its own cells, when he opened the first nut. Observing the scientist is inspiring the monkey's introspective response, but I would consider it unlikely that without the monkey having A Priori, of the experience, it would be able to relate to what it was seeing. To me the context would be a problem. But, what it *is* saying, is that the monkey had collected, stored, and was able to recall and embody, a previous state.

It is also as if the two of them, monkey and nut thief, became aligned in factor and notion. At the cellular level, each one (scientist and monkey) represent two arms of a tuning fork. Each experiencing their own internal harmonic and sharing the experience. This is an important point, not only in occasion to the map, but in context to the harmonics between states. We know that electrons and other particles passed around form harmonic relationships across space. In fact harmonics and interference, between states, would seem like a necessary bridge, which we'll look at again later. For now we have the map.

Chapter Nine

The Holographic Universe

In the Summer of 1999, while visiting some friends in Santa Barbara, I picked up the book, by the same name as this chapter, from an author named, Michael Talbot. Within its pages I found the template I was looking for. One that answered questions a lot better than the current cosmology was able to. It answered ever thing I could think of and more, but for starts, let's break it down so we're on the same page. And, even though it answered most of my questions, it failed to explain energy, as quantum mechanics and physics have. So let's take a look at the aspects that fit before we try and solve any conditions that need enhancement.[23]

The model was born in the minds of two men, one a physicist, from the University of London, David Bohm and the other a neurophysiologist named Karl Pilbram. Both were looking for a better model that could explain different phenomena more effectively, each in their own field of study. The framework most effectual for this aspect of our journey is the model which David Bohm developed to solve various puzzles that weren't being answered with the current model of physics.[24][25]

The holographic model is based on the working model of

a holograph. A holograph is a three-dimensional image of an object, created by two or more lasers, their light colliding in the air to reveal interference patterns within the fields of interacting light. Michael Talbot uses the scene from Star Wars; the moment when Luke first observes Princess Leia, streaming from R2D2's front panel, to give you a visual idea of its operation. "Help us Obiwan, your our only hope."[26][27]

How it actually works is, an object is photographed using a stream of light to detect and record the texture of the object, three-dimensionally. The shape and texture, of the object being recorded, adds an inference to the waves of light. This light is then merged into a pure stream of light and any interference, (or the difference) between them is recorded on a plate. The recorded image it captures looks nothing like the object it represents, which is an interesting trait about the actual waves of information that make up the quantum field, so to speak. This also creates an inherent problem, if you are trying to see it as forms, or even relative values based on the form, verses the information. You will either be looking at the information as waves; or energy and form. But, You can never see both simultaneously, because one is the information about the other. There is more to this dualistic nature, within the wave itself. It is not an uncertainty, but rather a rule. If you are looking at form, you are looking at the holographic image, and if it's a wave, it's the information.

The recorded image looks a lot like a smooth, glassy black lake, with a sudden disruption from thousands of tiny drops hitting the surface. This impression, even though looking nothing like the original object, has all the information about its texture, shape and color. When we divide a pure light to reflect, on this recorded image plate, and collide it back with the pure stream of light, the interference reveals to us, a three-dimensional image of the object we recorded.

With the actual holographic plate, it has the nature to hold a complete state of the information, uniformly, throughout the wave pattern. If we were to cut the holographic pattern, (on the plate) in half, we would still see the entire image reflected, even though the image intensity would be only half the original. But, the entire image is within every part of the wave patterns. This would explain how every atomic state is able to hold an instance of the global map, of the entire state, within it. In essence, every atomic state holds a piece of the context itself, as an anchor to the whole. This is the underlying texture of the universal fabric. It is what holds everything in stasis, based on its own referential location in a relative collection. Gravity, as Einstein depicts as this cause, in his theory of general relativity, is merely one of those fields.

Part Four

The Single Wave Model

Chapter Ten

Crystal, Chroma, and C

The holographic theory of the universe gives us a template, to wrap the phenomenal events, that makes up the universe. A smart fabric, so to speak. Within it is all of the information, we need, to decode our environment. We know that quantum mechanics tries to see everything in states of energy to mass relationships, but I believe it to be an entire layer of perception. We would describe it as a field of energy. In my opinion, quantum mechanics is completely independent from the texture of phenomena and need to be seen separately in order to be fully understood.

The Road to Chroma

My first interest into physics started when I was pretty young. In third grade, our class was invited to another class, to watch the Apollo 11, lunar lander, touchdown on the moon.[28] I was initially excited because I was infatuated with a little girl, in that class, and thought it was a great chance to play with her. Well, actually I was just hoping to talk to her, but something happened I wasn't expecting, at all. Call it the birth of Geek...

It started the moment the dust began to fly around the feet

of the landing module, and from that moment, I couldn't remove myself from watching the flickering TV, the whole time realizing that this was happening 15 minutes before we were seeing it. If you think about this for a second, these monkeys got into a tin can and just used a bunch of dynamic measurements and math, to get it to the efin moon. It was the logic that impressed me, at that moment. The day before I had just learned how to multiply with a zero, and today I was watching something happen, that was using the same logic, to accomplish the impossible. I realized that day that logic and math were an extremely powerful set of tools, and there was a lot, out there, still to be discovered.

Life is but a Dream

The next jolt I got was a few years later, in sixth grade. A hippy teacher I had, had us rewrite, "Row Row Row Yer Boat", and instead of the existing words, we had to replace them with words that meant the same thing. When we sang the song, the last line wasn't, "...life is but a dream", but came out, "...existence is but an illusion." Just after the end of that semester, we traveled to Washington DC and I saw the Apollo 11 landing capsule, up close, and the words hit me again. Existence is but an illusion, but there it was, right in front of me. It looked so real in person, and, when I was watching it on TV, it seemed real, but it felt like an Dream. The feeling that life was temporal and sensual never left me after that.

On the trip back, we stopped at a cave somewhere in the south, near New Mexico. We took a tour of a cave, following a park guide, deep into the Earth. At one point we reached a large room that could accommodate the entire group, which was about 30 people. There idea was, to show us that there was no dust in the cave. So, they killed the lights, and the guide shun a huge flashlight up into the tube above us, and guess what happened...? Absolutely nothing! There wasn't even so much as a glow evident in the entire room. I could see that the light was on. It was glowing red, slightly around the edge, but none of it's beam was either visible or had any effect on the environment of the room we were in. It was as if it was another illusion, I was standing here looking at...

Chroma

It wasn't until late 1989 that someone randomly told me that chroma was what removed the colors from light, and made the color green, like in a leaf. I had never heard of chroma, other than a "chrome" bumper, which framed the description in my mind as having something to do with metal.

Within a few days, someone else hit on the topic that, the reason a lead crystal glass will shatter when you strike a perfect C is that, the lead gets excited by the C note and begins to move, harmonically, with the frequency. The harmonic forces the glass to, **get out the way... Move!** And for the next few days I could hardly speak. It was as if my brain had been replaced by a machine and now I was driving, rather than riding. That's the best way I can explain it. Is life a dream?

What I realized was the significant relationship, that existed, between the C frequency and Chroma. I had already been pondering the relationships between sound and color, long before this, so hearing about chroma was a big jump from colors to energies and back. The idea I see is this; that each color of the light spectrum has a strong harmonic relationship to the notes of the musical scale. I saw red as an A frequency, orange as B, etc. all the way to G being violet. But this new information fit as well. I could see it. Chroma had to be relative to a C frequency, in some way. It had to be an energy, or an energy field, that was responsible for reflecting light and color.

Later, in 1994, I had a glimpse into how chroma actually worked. I could see that it was a tetrahedron, having four triangular sides. It had two rotational spin directions and four different processes. The first side was a clear window that allowed a photon to pass within the tetrahedron. Inside that, was a clear fluid state that smeared the photon out into the spectrum. The momentum of the rotation and the symmetry of the dual spin, were such that, one of the four sides passing by, caught all the removed values, and the last side met the remainder with a mirror, to reflect back the color an observer would see. And lastly, when the color passed (back) out, through the window, it could be diffused giving it texture, so to speak.

Fluidity

We see it happen whenever any matter acts like a liquid. Almost every element can be liquidized through temperature, and some can be influenced to act as if, when they are not actually in a liquid state, through vibration. A few days after I first arrived in Hawaii, an 8.2 magnitude earthquake hit Kobe Japan.[29] I remember that, because this particular morning, I woke up to tsunami sirens going off. The quake, which happened one year to the day after the Northridge earthquake in California, lasted almost a full minute. Most of the damage, caused by the quake, was to homes. The reason they said the damage was so widespread, wasn't from the magnitude, but the frequency. The frequency caused the earth (mostly sand) to actually move like water, causing the homes to literally start flowing along with the sand. I doubt anyone had a tuner handy at the time, but I'd bet it hit a D frequency.

When ever we think of fluidity, we usually think of it like water, but as a field, it represents a much larger part of the picture. It is likely that everything is harmonically bound to an energy field, including space. To me it seems completely impossible for something to float, without a medium. It is likely that space is actually a field of vibration, or more specifically a state of something, which solves a few fundamental problems that we'll discuss in a moment.

Chapter Eleven

Heat and Entropy

When we consider heat, we usually think of fire; an energy we use to warm water, heat our house, or cook food. It is such an indispensable tool, we use to do so many things, that it would take an entire encyclopedia to list the variances entirely. We were told, in science, that heat was when atoms get excited, but this is incomplete at best. They told us in physics and chemistry that, heat represents the energy loss, as entropy, in a dynamic system. While I agree to a degree, heat in my opinion is simply; the cost of any change, in any state, and it is governed by the laws of thermodynamics.[30]

The second of those laws, one which governs entropy, gives us a specific problem, in certain environments, called non-equilibrium states. The 2nd law, basically says, you can never have more energy than you start with, and that any system which converts heat to work has a unrecoverable loss, due to the exchange, called entropy.

The problem we have is similar to Alice's; tumbling down the rabbit's hole. We have no context for equilibrium in a vacuum. If the void of space is truly an empty vacuum, then there is no way for heat to dissipate from whatever object it is currently bound to. You could say that the energy would just continue off into the void, without a way to conduce the heat, thus keeping the Second Law true. But honestly, that seems to be hiding that the heat is being destroyed, in this particular model, when we know its impossible. Energy can neither be created nor destroyed, the first law of thermodynamics.

I do not believe that one, or the other, of these laws are coming into conflict with the other. I believe that they are both correct. And, I also do not believe that the entropy is escaping our scrutiny, in any form, partly in fact that I wish to see it, if it be, and this alone would constitute their becoming visible, as David Bohm observed. Another conflict, to this idea, would be a universe filled with huge heat and wind storms, from all the gathering waves, as they come in contact with any matter, while traveling through space. (hydrogen, helium, etc.)

It is far more likely space isn't empty. It is not a medium-less vacuum, but rather a medium that can not only absorb the dissipating radiations, but actually requires it as cost for any change in state. In fact, it is likely that without this cost there would be no need for entropy at all. In essence, it is part of the inertial behavior. Inertia, is simply when there is no change in a state, it will remain in that state; in motion or at rest, this is always the same.

If you consider the word "work" (in thermodynamics) as another cultural carry over, and simply replace it with the phrase, "manipulating the state". You can see that the other side of the equation, is heat. Now, lets have another look at our single wave. In it's current state, we would see it at rest and having near zero-entropy. But what happens if we introduce a second wave, which begins sharing electrons with the first, and the two form a weak bond, like a hydrogen lock? From this we will see an increase in heat.

So heat, in itself, is a state of whatever medium is bridging the two. This medium is shared between the two individual atomic states. We could simply look at this as a metaphorical cost for any change in state. But, we can again see the need for a conduit, a medium which can hold a state of temperature (in space), and act as the universal bank for spent heat. In essence, a neutral "ground". And, not just a correction to the mathematics...

Plasma - from the Greek words "everything form".

At one time, it was believed that everything was made of this unseen fabric, the Greeks referred to as plasma. Einstein referred to something he called aether, but I think he might have been referring to what was originally thought of as plasma, by the ancient Greeks. In Quantum Mechanics, they use a stop-gap, to fill this particular spacial void, using something named Hilbert Space. This is to account for the missing context, that can only be understood as necessary, in order for the math to jive. Little is actually known about plasma's relative relationship to mass. It is considered better understood in the visible range, involving tremendous amounts of heat. While the plasma itself is misunderstood.[31][32]

Modern plasma physics 101 starts off with the first sentence yielding to this "great unknown", by stating that its a "gas". Oddly enough, I see it as the "body" of heat, or any energy (that requires it), sometimes observed in a gaseous state, but... It is the body of the flame, but **not** the heat, the body of high temperature clouds of atoms and ions, but not the clouds of gas. But even more than that, it is merely visible at high temperature ranges, as a "glowing body of (whatever) energy". It seems that due to this visibility, the attributes have been applied to the container, and not the state. I see plasma as an entire field, and not only that, but I strongly see it as an obvious solution for the dissipating heat in space, or any non-equilibrium states. As it is not limited like Hilbert space would be, being bound to the elemental context, but extends out from that to the surrounding context, and likely bridges those states harmonically.

Chapter Twelve

The States of Plasma

To really understand what I mean, when I say plasma, we may need to define it, in a way, relative to the context we are considering, and form it in relation to the single wave model. So lets go back to the beginning (Nothing) and clarify that; the initial state, of perception observing nothing, is a single dimensional state, or more specifically, a *point*. This is the context for our primary perceptual state (universe). We will mostly refer to this as the universal perception. The next aspect of this contextual model is a two-dimensional fabric, I call the smart fabric. This is the *nature* of the plasmid body. It can be translated by an observer into any state. The other aspect of our model is the one observing the world of phenomena, which will refer to, mostly, as the observer.

The problem indicative to forming any kind of measure of plasma is, that it can only really be accomplished in agreeing to formulate a structure around one or various outcomes of its state. It cannot, as Einstein I'm sure realized, be measured, as it is a nature, and not a substance. It is the nature of the way the single wave oscillates. It is the nature of how energy flows across states. It is the nature of the body of energetic states. And, in no way, can it be considered a substance, and therefore cannot be measured.[33]

The wave acting in this state, can be broken down into two parts; The impulse of the wave, and the response of the wave. The response part of the wave, returns a counter-measure, so to speak, of the impulse part of the wave, through secondary harmonics in the form of inertia, or rather a natural resistance to change. Even more specifically, (the wave) is the resonant harmonic of the aligned states of perception, and gives the process not only its momentum, but also its rhythm and nature, by the very nature of the wave itself, which we'll delve into. And, as long as the state remains the same, there is no added cost to it.

The smart fabric must be viewed, conceptually, as a two-dimensional field, due to the purpose it fulfills to the observer perspective. It must represent the full scope of a complete perspective, perpendicular to the observer. It must contain the complete state and all states previous to it. The observer perception inflates this state, from the information embedded within the smart fabric, digesting it as a three-dimensional rendering, in a medium of the plasmid nature. This is the petri dish for the single wave model.

The plasmid characteristics are actually the character of the fabric nature being translated into behavior and traits, specifically the holographic representation, as a whole. We can not visually see the fabric, as it is. We can only see the holographic image it represents. Unless, like the slit test we're recording, while the context is missing, and only catch the waves.

The states of plasma

The fabric itself is a field. This field is effected by the states infused within it. There are seven states that make up the plethora phenomenal universe. There is a natural harmonic relationship between each of these seven states and a color, as well as a frequency. Spectrophotometry shows that all atoms refract a spectral array in specific bandwidths, that I believe have a direct correlation to this model as well. The frequencies, relational colors, and energies are listed below. It will take some time to work out the blueprints for all the elements, in the periodic table, and the way in which we map it out. Until then, we have the seven fields, which are...

1. A , red , heat
2. B , orange, gravity
3. C , yellow, chroma
4. D , green , fluidity
5. E , blue , electric
6. F , indigo , magnetic
7. G , violet, light

Each of these seven states are within every form of nature, for instance. If something is a gas, it is a mixture of heat effecting the fluidity. The scope of heat expands the state of fluidity and overloads the chemical bonds, to free a carbohydrate from wood, or separate water molecules into steam. And all of these are the effects of combining plasmid state parameters.

The reason the observer state is different from the core state, the point, is that it requires a different set of qualities to accomplish its purpose. It must bridge the gap between the single point and the inflated universe. The single point is totally adequate for crunching data. It really only needs a context, in which to operate, and has no spacial requirements beyond this. While the smart fabric is just that. It does require at least a two-dimensional context to hold and reflect the values called by the perception observing it. It does this by passing its information, harmonically, as an impression of the total state. The impression is literally held by the observer as A. Tension, which forms a harmonic bond, allowing all the information to pass seamlessly between them.

Chapter Thirteen

The Primary State

Ultimately we are faced with the reality that the world, and all the phenomena we experience, are really arrays of dynamic values in a holographic impression, expressed in the harmonic values of energies and form. In one arena, the information is contained within a non-spacial state, while the same information is cast as the harmonic resonant impression of a three-dimensional pattern. This is, in turn, translated by the observer into every shape, effect and sound, in the phenomenal world around us. We do this by paying attention to the global map, (the universal A. Priori), embedded within every atomic state. This information is passed through the smart fabric to the faculties of the (unique) perception observing it. And we dig it as an event.

But, something's happening here. If we consider the original context being, Nothing. We must conclude that, there could be only one single infinite state, and there can be only one single perception observing it. Otherwise, it would lead to an null exception; trying to recreate it in any way. There are no more resources for a second anything. Just the fact that the original context is whole, only allows us to contain parts of it, to complete the perception and state, in every element of the total state.

There are a few ways you could do this. One way would be something called boundaries, which are like containers, or by creating something referred to as an instance, in programming, of the original Single Wave. Boundaries would cause spacial consequences, while instancing requires none. An instance, theoretically, will have all of the parameter fields of the original, but will not include any of the previous state's value, being that it is an original of the primary state, and therefore null. To make a long explanation even shorter, each object has it's own unique state of perception, just as each perception has it's own unique state of total mass.[34]

This creates a very interesting secondary problem. The difference between each unique state can not be undone. They might as well exist in entirely different universes. The only way for these, fully separate, contaminants to communicate is through a medium that operates as a conduit between states, to pass the part of the information the two states need to share; electrons, heat, magnetism, gravity, spin, clock, and more. There are two ways this can happen, either via harmonic resonant; a value passed either directly or through a statistical global range, as a field. Or, through an impression of the harmonic, to an adjacent state, in the form of a wave.

Chapter Fourteen

The Atomic State

The atomic state is an array-set of values, that make up the information of the current state, of any atomic state. The atom, on the other hand, is the three-dimensional outcome when perception collides with that state. In other words, one is the information state (about the form) and the other is the form. The information always precedes the form, as the A. Priori suggests, prior to experience.

The atomic state is made up of seven (7) individual states that interact harmonically, to create the interference patterns for the forms they represent. There are three primary states and three secondary harmonic states, with the seventh state being light at its center. The first three primary states are electricity, fluidity, and chroma. These make up the electron, neutron, and proton respectively. The secondary harmonics are something very important! These are the anchors of inertia. Their values are based on the resistance to any change in state. These are magnetism, gravity, and heat. Gravity would represent the resistance to any change in momentum. Magnetism would result from the spin of the electron state momentum, and the heat results from any change in state. Together these make up the nature of inertia.

The Plasma Screen

Whenever I mention the two-dimensional surface, I am really talking about a representational value. The surface isn't really a surface. It merely represents a surface to the observer, and only in the way it presents its information. Its shape isn't important until we need to discuss it in a way it can be understood, similarly to establishing a context for the original Nothing. I also want to clarify that the screen isn't exactly like a TV plasma screen, but is similar enough to run with. The main difference is how it presents its information.

To begin with, lets look at the single wave, as our starting point. If we were observing it, in context with the smart fabric, we would see a single circular wave, oscillating with the clock. Within in it would be 7 layers, each representing one of the seven states. As the states interact with each other, we would see it as a single concentric oscillating wave, with variable interference patterns. These interference patterns make up the dimensional object, in a holographic state. In this case, a hydrogen atom.

Single Wave
[illustration]

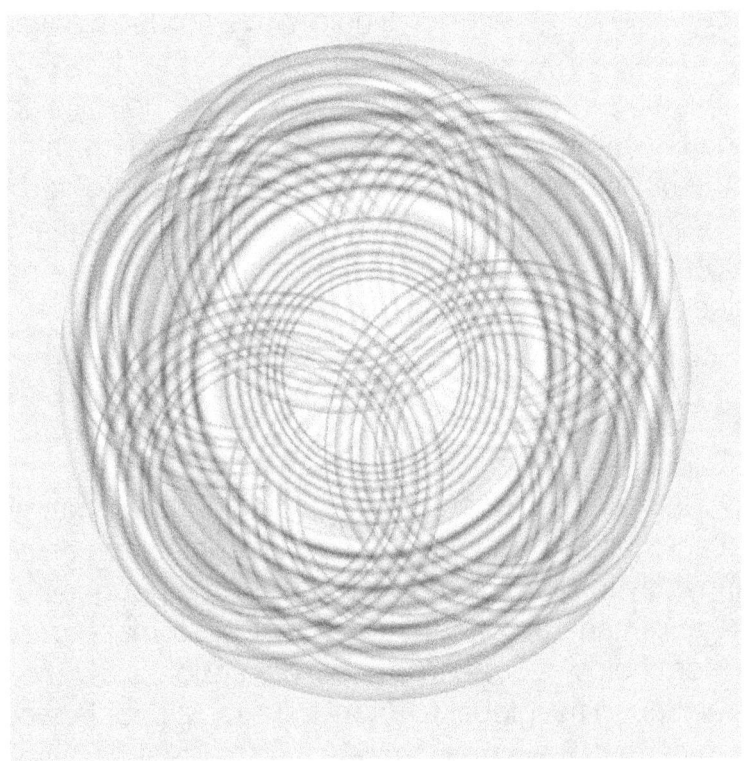

Tory

The first time I heard about Tories, I was hiking in the Rockies. A Tory is a stack of rocks, usually three, which helps hikers see which way the path, goes forward, when it is difficult to detect, I.E, rocky surface, or a river crossing. The word comes from, once again Greece, and describes the markers set in the corners, of a piece of land, to mark the boundaries. When used for land they are referred to as a Terra-Tory, Terra meaning earth. In Single Wave, a Tory is where three harmonics converge into a peak in the signal of the state. This is likely where the particles come into view, while it is being observed.

If we were to inspect our Single Wave Model, we would see multiple peaks, revolving around the wave, and converging into the harmonics that make up each Tory. Together these Tories represent the corners of our atom , in the holographic state. I believe that if we are to decode the value of states, as raw data, we will need to find a way to do it via inspection of the Tories and how these peak values can be assessed and understood.

The Big Screen

Now lets move back and fill in the rest of the picture with a scene; a little boy, in a park near a fountain, on a warm sunny day in July. You can almost smell the hot dogs. This image is made up of a huge number of atoms, and each of them holds values for position, dimension, and time, all in relation to the global A. Priori. I, like most people, would imagine that, if we were to view the smart fabric it would be filled with waves, right..? Actually, it's only going to be a single wave, with all the information to every state (within it), and all previous states, passed by time. Remember that each $1/10e43$ of a second, is the total length of a single state of total mass. So the depth in the image is, actually, in the past states moving toward us, by however long it takes light to reach us, from whatever depth were inquiring. In other words, perception/time = state of total mass.

Other things worth considering are some of the global affects, to the entire scene simultaneously, like the sun adding heat to everything. Or the moisture and carbon dioxide emissions from all the flora. Or... the billions of waves passing through the scene in the form of waves, in every bandwidth known, starting with cell phones. These values would be shooting in and around, and through, all the information it affects, even slightly. But the big one is beneath our feet, in the Earth itself. This is a global map index for all the existing values connected with it. Even in the air, through the thin fibers of gravity. And this is part of the solar system's map, which is part of the galactic map, and all these values are written into the entire environment through the individual pixels of the atom interface.

Chapter Infinity

God, the Universe and Nothing

God doesn't play dice...
God plays Craps

And the story is this. In the beginning was perception... I know it probably wasn't the God that the religious folks were hoping for, but it is what it is. The fact that there is no way to have anything without some kind of perception observing it also, is. Where perception came from, in the beginning, might be locked inside the single wave, waiting to be discovered. I'm sure that the information is there. It just might take a much more subtle approach to the data than we are currently able to offer. But, don't worry... It's bound to occur eventually.

The Big Wave

Once the single wave, in our original model, began to grow exponentially, there was an equal expansion of hydrogen growing in an infinite field. It would have begun to produce a lot of heat and static, quickly filling space. At a certain point the gravity would begin compacting the volume into enough pressure and in no time, we have a volatile mass. And, once the primary chain reaction of heat jumped the fence, it would lead to atomic linking and molecular development, not to mention galaxies, stars, and planets, through secondary reactions and aggregate materials. One of those aggregate results will eventually lead to life as we know it...

Everyone who was alive in 1996 probably remembers seeing Hale-Bopp pass through our solar system. It was headed past the Earth, to do a loop around the Sun, eventually coming back to shoot past the Earth again, before heading back out into space, in 97. It was so big and so close, you could see the thing in the middle of the day. NASA sent a satellite out to meet it, to see if we could gain more information about it. What we found was amazing. It gave us another clue to the Big Wave! It turns out, that there was ethyl-cyanide in the spectrophotometry of the tail, which mixed with (wet) water starts the process to making amino acids.[34]

It would appear that comets are the universal sperm impregnating planets, with the seeds of life, that are wet and warm enough to support algae. This would have taken, at least, 4-10 billion years to occur, but once it started it would have begun the road to forming the first observer states of perception. There would have been no knowledge of glow or warm, or green, and so the first algae wouldn't have had any photosynthesis. But, now there was a whole new set of information to explore, which needed better perceptual faculties. Each step gaining specializations and developing a better relationship with the environment and available elements.

The Stages of Light

Eventually, as we all know, through all the processes we have seen and explored, we had an explosion of human beings. Humans have rich developmental specialization skills, that as we have seen have both a curse and blessing they come with. As our story of human's possession of fire began, so does our evolution through these various stages, which represent the texture of our fields, another metaphor. Our evolution has been one of perception, as a global state. Each element holding a part of the map.

Our discovery of fire was likely around 30,000 years past, and it was this that woke the roots, the red, A frequency, of heat. Next was the second state, around 12-14,000 years ago. This began the era of farming and granaries, nourishment and sentient lives, like gravity to the Earth. B frequency, orange. The next big shift was through specializations, having more food than people. 8000 years ago. And this led to creative explorations into the fabric of the space, and metal. Metal leads to armies and Empires, to reflect like gods themselves, like chroma. C frequency, yellow. Most pop music is in the key of C.

Fluidity is curious. It would have been around 5000 years

ago, and some evidence points to a possible flood. The Great Barrier Reef, around Australia, started growing... Again, approximately 5200 years ago, and the oldest tree on earth is located in the peaks of the Sierra's in California, and chimes in as the same age at 5,200. D frequency, green. Another condition that seems to be part of this cycle is that the peaks are getting closer in frequency. Next comes the fun ones. From around 400-0 BCE, a few humans managed to change the earth a few times. Socrates, in the west, using words to break down the truth, while in the east we had the thousand schools of thought. Like the throat opening. Electric, E frequency, blue.

My favorite shift is the renaissance period through Newton. Approximately 100 years between 1500 and 1600. This is the mind waking up to depth of field, and the laws of motion. I sometimes wonder, looking out over the open ocean, why they couldn't see the curve in the Earth previous to this time. But, it was happening all over, at that time. It was like the world suddenly took shape and became a rich texture of behavior. And bodies moving about in the space above them. It had so many patterns to discover. Magnetism, F frequency, indigo.

The Here and Now

Throughout the last 500 years, the world has come around to become a technological maze of everythingness. Having reached what appears to be another peak in the frequency of change, we are faced with a growing number of conditions that may, or may not, determine our fate. What is certain is that a bit of the character, of this state of perception, has come through the use, and engaging, that occurs across the internet. It has shown us, as Jeremy

Rifkin points out in his video, to be a highly empathetic beings. It is the nature of this same behavior among tribes, when they need each other. The internet has given us an infinite collection of information that only exists as values in stasis, waiting to be inflated. A metaphor.

Bibliography

Chapter 1

[1] background cosmic radiation - The accidental discovery of CMB in 1964 by American radio astronomers Arno Penzias and Robert Wilson was the culmination of work initiated in the 1940s, and earned the discoverers the 1978 Nobel Prize.
http://en.wikipedia.org/wiki/Cosmic_background_radiation

[2] Special Relativity - It was originally proposed in 1905 by Albert Einsteinin the paper "On the Electrodynamics of Moving Bodies".
http://en.wikipedia.org/wiki/Special_relativity

[3] Georges Lemaître Big Bang Theory - 17 July 1894 – 20 June 1966) was a Belgian priest, astronomer and professor of physics at the French section of the Catholic University of Leuven.
http://en.wikipedia.org/wiki/Georges_Lemaître

[4] Edwin Hubble - Hubble is known for showing that the recessional velocity of a galaxy increases with its distance from the earth, implying the Universe is expanding.
http://en.wikipedia.org/wiki/Edwin_Hubble

[5] Dark Matter - Dark matter is one of the greatest mysteries in modern astrophysics. It cannot be seen directly with telescopes; evidently it neither emits nor absorbs light or other electromagnetic radiation at any significant level.
http://en.wikipedia.org/wiki/Dark_matter

[6] Gravaton and String Theory – Some of the Grand Unification Theories.
http://en.wikipedia.org/wiki/Grand_Unified_Theory

[7] Hawaii infra-red star distance analysis - Using the Infrared Array Camera (IRAC) aboard NASA's Spitzer Space Telescope, astronomers have detected about a dozen very red galaxies at a distance of 10 to12 billion light years from Earth (cfa Harvard 2005). According to the Big Bang model, these galaxies existed when the universe was only about 1/5 of its present age of 13.75 billion years. The unpredicted existence of "red and dead" galaxies so early in the universe challenges Big Bang theories relating to galaxy formation (cfa Harvard 2005). Analysis show that galaxies exhibit a large range of properties. Young galaxies with and without lots of dust, and old galaxies with and without dust. There is as much variety in the so called "early universe" as we see around "today" in galaxies closer to Earth.
http://www.dailygalaxy.com/

Chapter 2

[8] Uncertainty Principle - Introduced first in 1927, by the German physicist Werner Heisenberg, it states that the more precisely the position of some particle is determined, the less precisely its momentum can be known, and vice versa.
http://en.wikipedia.org/wiki/Uncertainty_principle

[9] Double Slit Test – The double-slit experiment is a demonstration that light and matter can display characteristics of both classically defined waves and particles; moreover, it displays the fundamentally probabilistic nature of quantum mechanical phenomena.
http://en.wikipedia.org/wiki/Double-slit_experiment

[10] Bose-Einstein Condensation
http://en.wikipedia.org/wiki/Bose–Einstein_condensate

Chapter 3

[11] Niels Bohr's - Atomic Model
http://en.wikipedia.org/wiki/Bohr_model

[12] Amedeo Avogadro – Molecular Model
http://en.wikipedia.org/wiki/Amedeo_Avogadro

Chapter 8

[15] Higgs Bozon
http://en.wikipedia.org/wiki/Higgs_boson

[16] Newton – Laws of Motion
http://en.wikipedia.org/wiki/Newton%27s_laws_of_motion

[18] EPR Paradox
http://en.wikipedia.org/wiki/Newton%27s_laws_of_motion

[19] Copenhagen Quantum Mechanics
http://en.wikipedia.org/wiki/Copenhagen_interpretation

[20] Quantum Mechanics
http://en.wikipedia.org/wiki/Quantum_mechanics

Chapter 9

[19] see chapter 8

[21] Emmanuel Kant - A Priori
http://en.wikipedia.org/wiki/Immanuel_Kant

[22] Jeremy Rifkin - Empathic Civilization
https://www.youtube.com/watch?v=l7AWnfFRc7g

Chapter 10

[23] Michael Talbot - Holographic Universe
http://en.wikipedia.org/wiki/Michael_
Talbot_%28author%29

[24] Karl Pilbram
http://en.wikipedia.org/wiki/Karl_H._Pribram

[25] David Bohm
http://en.wikipedia.org/wiki/David_Bohm

[26] Star Wars 1976

http://en.wikipedia.org/wiki/Star_Wars

[27] Holography
http://en.wikipedia.org/wiki/Holography

Chapter 11

[28] NASA Apollo 11 Mission
http://www.nasa.gov/mission_pages/apollo/missions/apollo11.html

[29] Kobe Earthquake sand flowing like water
http://factsanddetails.com/japan/cat26/sub160/item866.html

Chapter 12

[30] Thermodynamics
http://en.wikipedia.org/wiki/Thermodynamics

[31] Plasma
http://en.wikipedia.org/wiki/Plasma_%28physics%29

[32] Hilbert space
http://en.wikipedia.org/wiki/Hilbert_space

Chapter 13

[33] Einstein – aether
http://en.wikipedia.org/wiki/Einstein_aether_theory

[34] Java Programming – instance
http://docs.oracle.com/javase/tutorial/java/javaOO/
classvars.html

Chapter 16

[35] Hale-Bopp - white paper on spectrophotometry
http://iopscience.iop.org/0004-637X/689/1/613/
pdf/75253.web.pdf